PRIMARY POINTS OF DEPARTURE

*selected and rewritten by Jenny Murray
from Points of Departure 1 - 4*

Originally compiled by members of the
Association of Teachers of Mathematics

INTRODUCTION

This book contains over 70 starting points from *Points of Departure* 1 - 4 selected for pupils at Key Stage 2, and those working at similar levels. The chosen items have occasionally been extended, sometimes slightly altered, and frequently shortened, with extension ideas put in the notes at the end of the booklet. Some starting points, such as 9 (SIXES), 11 (PARTITIONS), 16 (PICKING STONES), 40 (ELEPHANT WALK), 49 (NUMBER PATTERNS) and 57 (QUINCUNX) may also be found useful for Year 2 pupils.

Teachers can introduce the starting points as they stand to individuals, pairs or small groups. However, this is not the main intention. Rather, the book should be seen as a resource of ideas that teachers will wish to adapt and to present in ways with which they, and their pupils, feel most comfortable. Some, for example, 24 (HALVING THE BOARD), 37 (ROUND AND ROUND) and 63 (EQUABLE TRIANGLES), could well be used for initiating a discussion which leads to the given diagrams being put on the black (or white) board. In this case the words given in the starting point can be adapted as an introduction to a class lesson. Some of the starting points require equipment such as interlocking cubes, squared or dotty paper. Sometimes, especially with younger pupils, other pieces of apparatus will be found useful.

However the points of departure are to be used, teachers are strongly advised to spend a few minutes exploring the ideas before using the points of departure with a class.

Some of the starting points may only fill a short time while others could occupy pupils all week. As all classes differ it is not possible to indicate which of these is which! Some items, such as 9 (SIXES), could be picked up and used to fill odd moments.

Most of the starting points can be attempted over a range of abilities, allowing for differentiation. For example, the numbers in 22 (ARITHMOGONS) and 26 (TABLES) can be made easy, and 7 (ALL DIFFERENT) and 17 (MAX BOX) kept very practical for some pupils. At the same time other more able pupils may be attempting the extension ideas.

The chosen starting points range over various areas of maths in the National Curriculum, but all have *Using and Applying Mathematics* in mind.

Many of the starting points allow opportunities for pupils to select, not only their own mathematics, but also the materials with which to work.

All the starting points can be used to develop the use of mathematical language. There are cases for pupils to try predict the outcome, to draw the next example, to suggest another way of tackling the problem. Often there is a need to attempt to formulate rules.

The extension ideas given can be used in a number of ways, or not at all. Many pupils will think of their own extensions. "What happens if I change this?" This is good. In any case the extensions are only intended for more able pupils.

CONTENTS

1. ROUTES
2. RAILS
3. DOTS AND LINES
4. MULTIPLICATION SQUARE
5. FAULT LINES
6. DOMINOES
7. ALL DIFFERENT
8. 1089
9. SIXES
10. SQUARES IN SQUARES
11. PARTITIONS
12. CUBE NETS
13. PALINDROMES
14. SUBTRACTION PATTERNS
15. CHAINS
16. PICKING STONES
17. MAX BOX
18. FOUR FOURS
19. PAINTED CUBES
20. CONSECUTIVE SUMS
21. SETS OF FIVE
22. ARITHMOGONS
23. MAGIC SHAPES
24. HALVING THE BOARD
25. HAPPY NUMBERS
26. TABLES
27. MOVING TESSELLATIONS
28. HICCUP NUMBERS
29. RECTANGLE AREAS
30. STRIPS OF SQUARES
31. JUGS
32. POLYGON SYMMETRIES
33. HEXIAMONDS
34. DOTTY SHAPES
35. FINDING TRIANGLES
36. ADDING DIGITS
37. ROUND AND ROUND
38. ALL THE DIGITS
39. AFRICAN NETWORK PATTERNS
40. ELEPHANT WALK
41. PALINDROMIC DATES
42. DOMINO ARITHMETIC
43. CALENDARS
44. TOTAL'S
45. PATIO PATHS
46. THE TETHERED GOAT
47. COUNTING TRIANGLES
48. TRAVELLING SALESPERSON
49. NUMBER PATTERNS
50. SUMS AND PRODUCTS
51. TILINGS
52. PROJECTIONS
53. STICKS
54. RED AND YELLOW
55. ROUND THE BLOCK
56. SUMS OF DIVISORS
57. QUINCUNX
58. STIFF LITTLE FINGERS
59. 10 X 12 or 11 X 11
60. FIRST WITH THE FACTORS
61. PATIO TILES
62. LINES AND REGIONS
63. EQUABLE RECTANGLES
64. STICKY TRIANGLES
65. COLOURED CUBES
66. THE GAME OF 25
67. A4
68. DIFFERENCING
69. NOUGHTS AND CROSSES
70. FOLDING STAMPS
71. CUBOIDS
72. HOW GREAT?
73. BRICKS

AT THE END: Some notes and extension ideas

1. ROUTES

Start at A and travel along lines only in these two directions

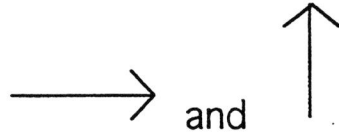 and

In how many different ways can you get from A to each of the lettered points?

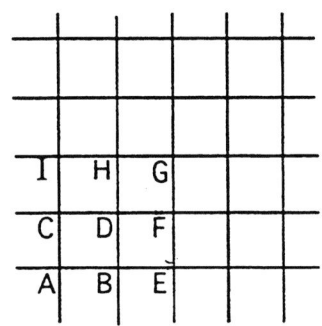

Try some other points.

Can you spot some patterns?
Can you generalise them?
Can you explain them?

Start at A and travel along lines only in these directions:

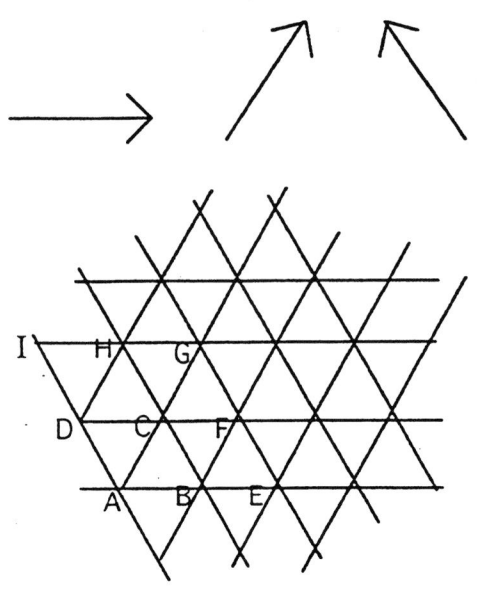

Investigate as before.

2. RAILS

A child has a large number of curved rails in a train set. They are all quarter circles and can be placed either

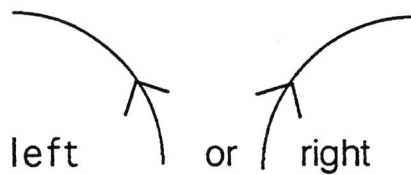

Investigate ways to make closed tracks.

3. DOTS AND LINES

Mark six dots on a sheet of plain paper.

Straight lines go through every dot.
How many are needed?

Try for other arrangements of six dots.

What is the maximum that might be needed with only six dots?

What number of lines smaller than the maximum can you make?

4. MULTIPLICATION SQUARE

Make a ten by ten multiplication square.

Investigate the number patterns in it.

5. FAULT LINES

Here is a rectangle made with some dominoes.

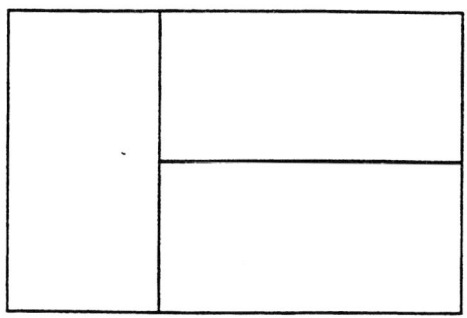

Each domino is a 2 X 1 rectangle.

A fault line is a straight line joining opposite sides of the rectangle.
So these rectangles have fault lines marked by the arrows.

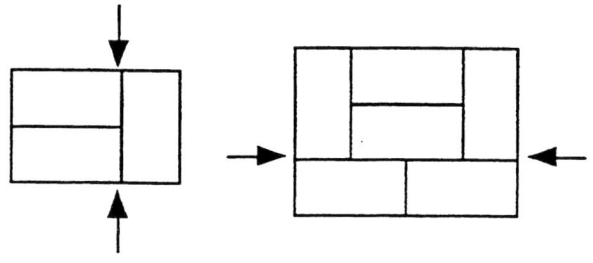

What is the smallest fault-free rectangle you can find?

What is the smallest fault-free square?

6. DOMINOES

Dominoes are put together in the usual way.

Investigate the possibility of forming closed chains.

7. ALL DIFFERENT

Throw three dice.

If two or more dice show the same value, throw these dice again. Keep throwing until all three dice show different values.

What is the average number of throws needed to get all the dice showing different values?

Try for four dice, or five, or six.

8. 1089

Write down a 3-digit number,

say,	742
Reverse the digits	247
Subtract	495
Reverse the digits	594
Add	1089

What happens if you start with 564?

Do you always get 1089?

9. SIXES

What sums can you find with the answer six?

10. SQUARES IN SQUARES

How many squares here?

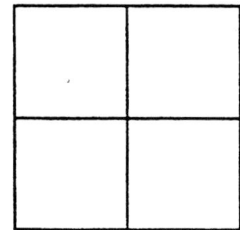

Five! Where are they?

How many squares
        on a 3 X 3 board?
        on a 4 X 4 board?

What about rectangular boards?

How many squares here?

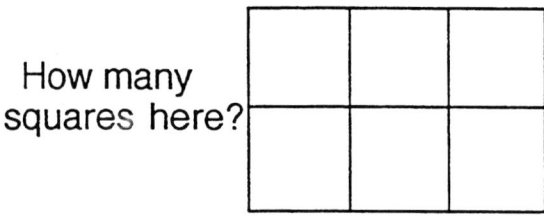

11. PARTITIONS

You could use Cuisenaire rods to help with this problem.

The 3-rod can be partitioned (split up) in four different ways:

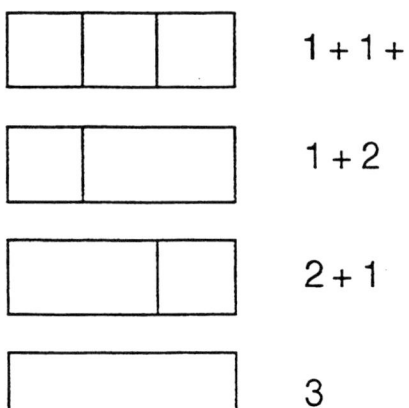

1 + 1 + 1

1 + 2

2 + 1

3

Investigate partitions of different numbers.

12. CUBE NETS

This diagram shows one possible net for a cube.

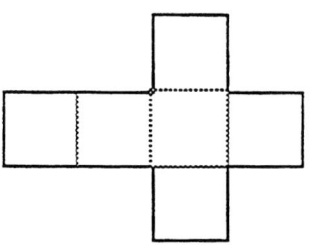

How many different cube nets can you find?

13. PALINDROMES

Choose any number,   2 1 6
reverse the digits   6 1 2
and add: 8 2 8

8 2 8 is a palindromic number (the same backwards as forwards)

Try another number   1 5 4
reverse and add: 4 5 1
                         6 0 5

6 0 5 is not a palindromic number, so we repeat the process:

                        6 0 5
                        5 0 6
                    1 1 1 1

1 1 1 1 is palindromic.

Does this always happen?

Investigate.

14. SUBTRACTION PATTERNS

This is a subtraction pattern:

```
    5   11   8   15
      6   3   7   10
        3   4   3   4
          1   1   1   1
            0   0   0   0
```

it starts with the four numbers   5   11   8   15

The next row is obtained by working out the differences between numbers next to each other in the row before - we imagine that the last number in the row is next to the first one.

The pattern stops when the numbers are all zero.

Investigate for different sets of starting numbers.

15. CHAINS

$$6 \rightarrow 3 \rightarrow 10 \rightarrow 5 \rightarrow 16 \rightarrow \ldots$$

Rules:
1. If a number is <u>even</u>, divide by 2.
2. 1. If a number is <u>odd</u>, multiply it by 3 and add 1.

Continue the chain above. What happens?

Choose other starting numbers and see what happens.

16. PICKING STONES

This is an old Chinese game for two players.

They take it in turns to select stones from two piles by taking

   either: at least one stone from one of the piles
             (all of them if you like)

   or: the same number of stones (at least one) from each pile.

Players who take all the stones on their turn are the winners.

Investigate for different numbers of stones.

17. MAX BOX

Suppose you have a square sheet of card measuring 15cm by 15cm and you want to use it to make a box (without a lid).

You could do this by cutting squares out at the corners and then folding up the sides.

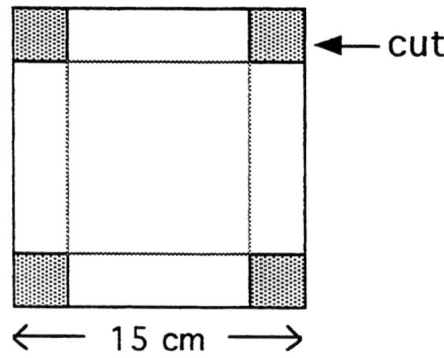

Suppose you want the box to have the maximum volume.

What size corners would you cut out?

18. FOUR FOURS

What numbers can you make using four 4's and mathematical symbols?

For example:

$$\frac{4}{4} + \frac{4}{4} = 2$$

$$\frac{4+4}{4} + 4 = 6$$

Find some you can't make.

19. PAINTED CUBES

A three-by-three cube is made out of little blocks.

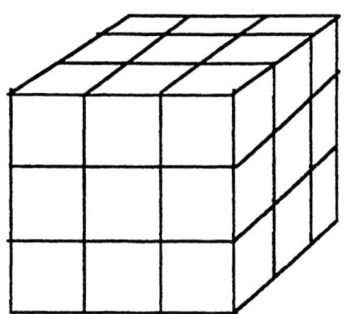

The outside is painted red.

How many little blocks have
    3 sides painted?
    2 sides?
    1 side?
    0 sides?

20. CONSECUTIVE SUMS

$$15 = 7 + 8$$

$$9 = 2 + 3 + 4 \text{ or } 4 + 5$$

$$10 = 1 + 2 + 3 + 4$$

These three numbers can be written as the sum of two or more consecutive integers.

See what numbers you can make.

Which numbers cannot be made like this?

Which numbers, like 9 above, can be split up in more than one way?

21. SETS OF FIVE

Here is a set of five numbers.

$$\{1, 2, 3, 7, 12\}$$

17 can be made by adding some of them together:

$$17 = 12 + 3 + 2$$

Can you make 11? 23? 25?

(You are allowed to use each number only once.)

What is the highest number you can make?
Which numbers can't be made?

Try other sets of numbers.

22. ARITHMOGONS

In an arithmogon, the number in the square must be the sum of the numbers on either side:

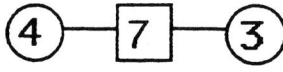

Solve these arithmogons:

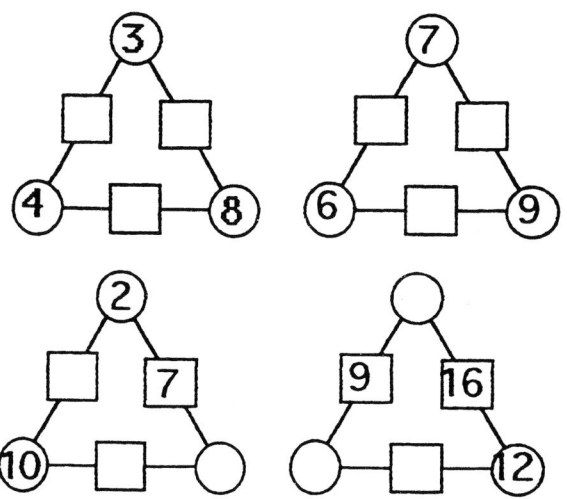

Make up some of your own.

23. MAGIC SHAPES

For the magic square you can put the numbers from 1 to 9 in the spaces so that each line adds up to the same total:

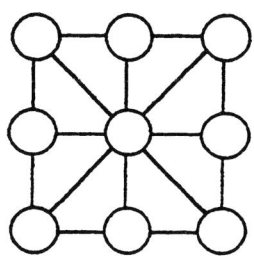

Try these magic shapes too:

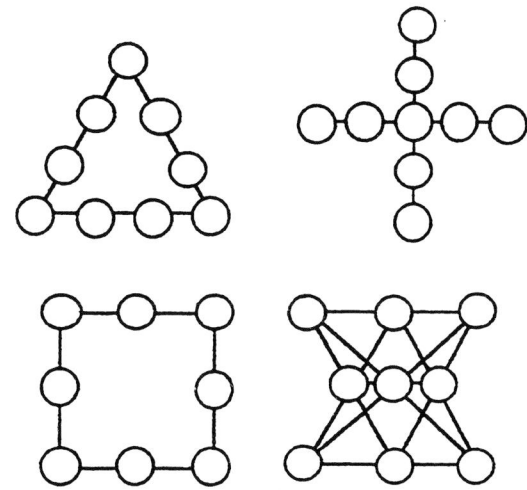

24. HALVING THE BOARD

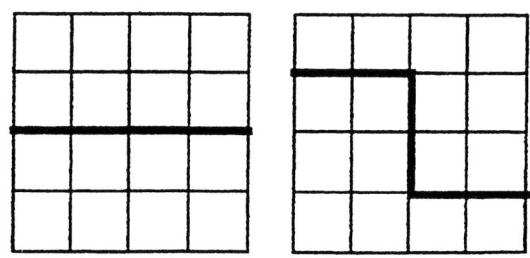

Here are 2 ways of cutting a 4 x 4 board into 2 identical pieces.

What other ways are there?

25. HAPPY NUMBERS

23 is happy because:

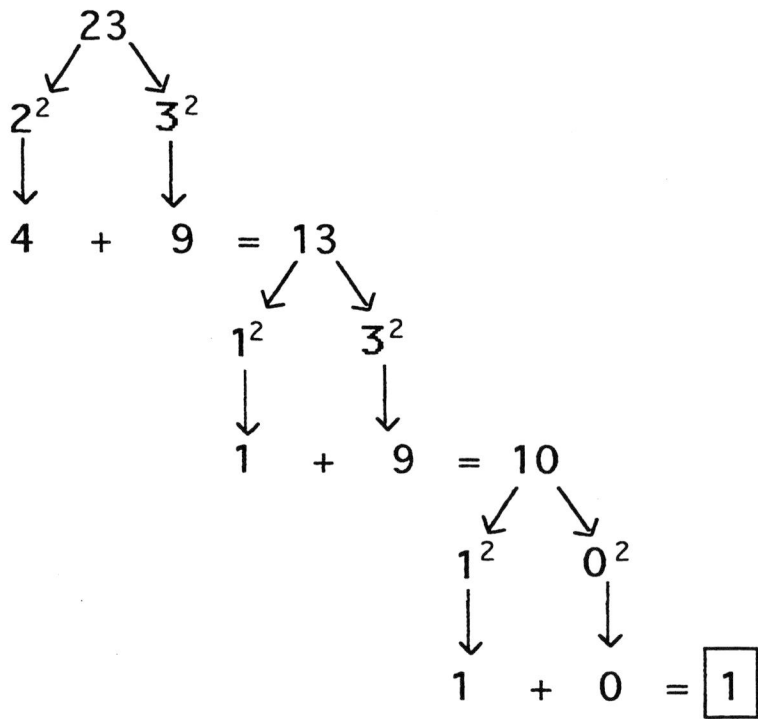

If you end up with a 1, the number you started with is happy.

Is 15 happy? What about 7? or 24? Try some others.

How many numbers less than 60 are happy?

26. TABLES

These are examples of 2 by 2 tables:

+	3	5
2	5	7
8	11	13

×	3	8
2	6	16
9	27	72

−	3	8
2	1	6
9	6	1

These are incomplete 2 by 2 tables. Can you complete them?

×		
	10	15
	14	21

	11	5
21	10	16
13	2	8

+	2	3
		30

27. MOVING TESSELLATIONS

A common tessellation can be transformed in a variety of ways. One way is to imagine the shapes all moving apart (outwards). The gaps between the shapes can then be filled in various ways, depending on how far apart they are, and how they are oriented relative to each other.

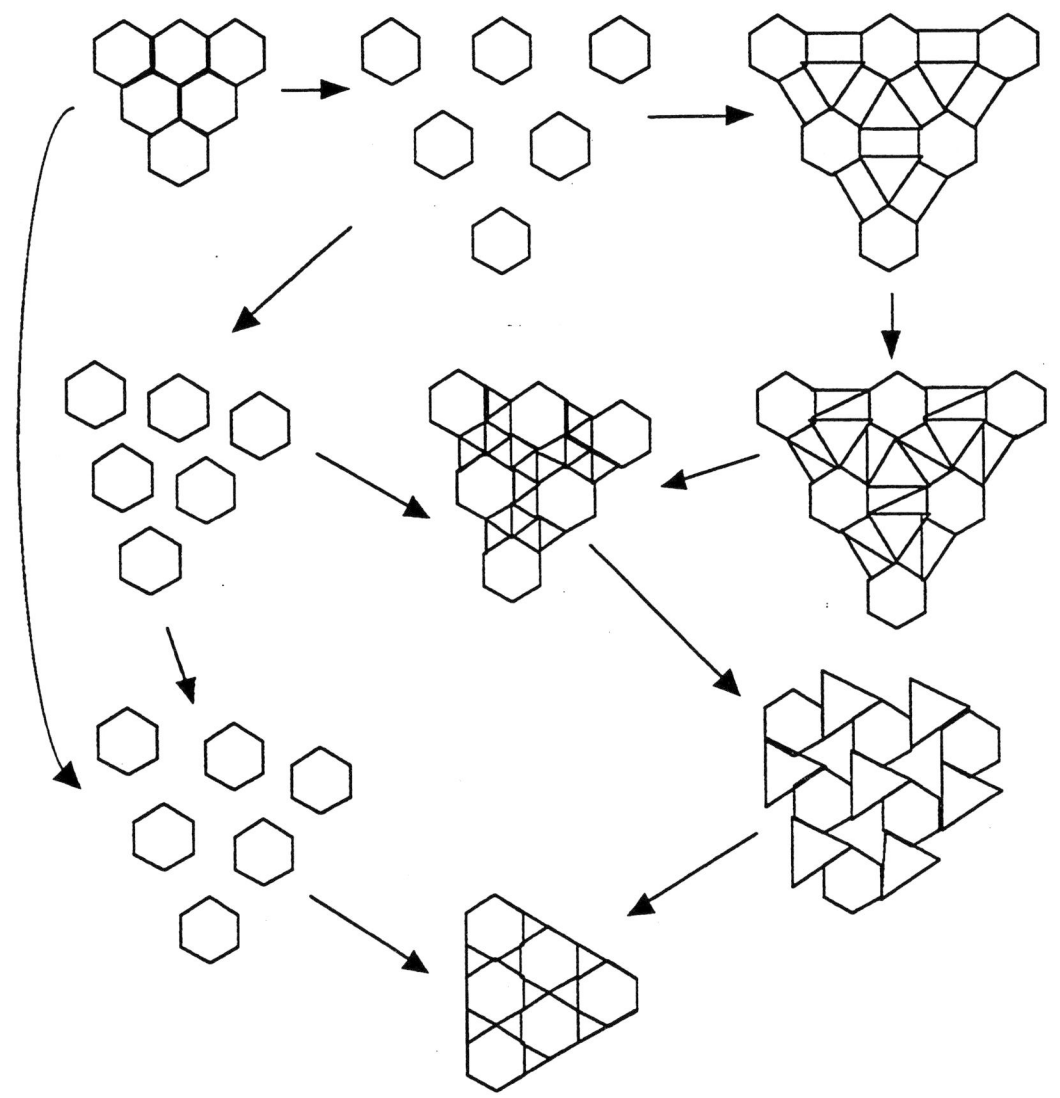

What happens when we start with other tessellations -
triangles, squares .... ?

28. HICCUP NUMBERS

Choose a 3-digit number, say 327, and repeat it, 327327.

Divide the number by 11, by 13, and by 7 - what happens?

Investigate other 'hiccup' numbers in this way.

29. RECTANGLE AREAS

Use squared dotty paper. Here are some rectangles of area 10 square units.

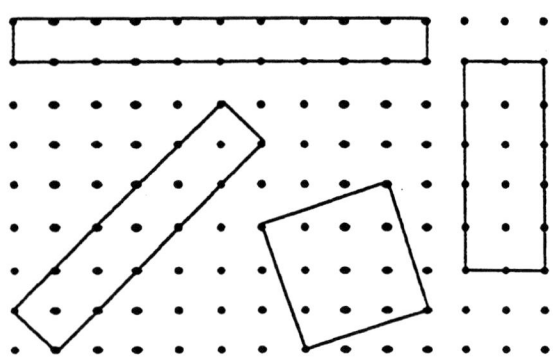

Can you find more?

Try other areas.

30. STRIPS OF SQUARES

Strips of squares are made and coloured in two colours.

We call the shapes equivalent if they can be reflected or rotated into one another.

How many different strips of length five squares can you make?

31. JUGS

If you had a 3-litre jug and a 5-litre jug, how could you use them to measure 4 litres?

Make up some more problems like this.

32. POLYGON SYMMETRIES

Quadrilaterals can have

no lines of symmetry

or 1 line

or 2 lines

or 4 lines

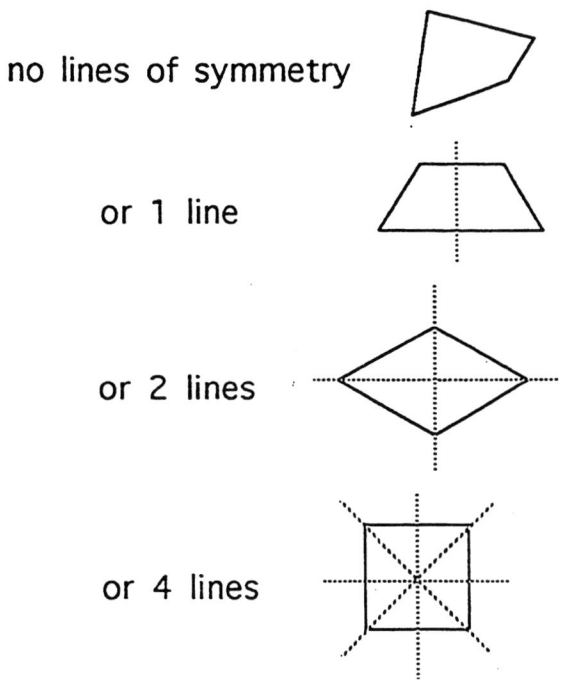

How many lines of symmetry can triangles have?

33. HEXIAMONDS

Hexiamonds are shapes made from six equilateral triangles.

Here are some:

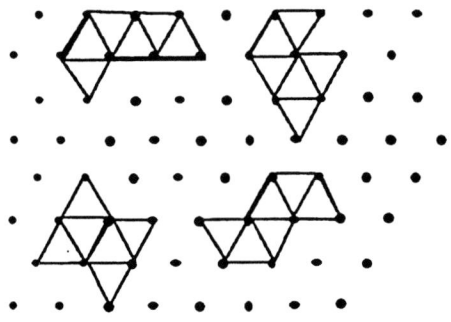

How many more can you make?

34. DOTTY SHAPES

Make some shapes with no dots inside:

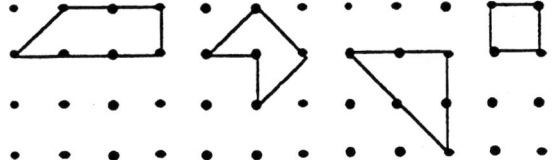

Find the area of each shape and the number of dots on its perimeter.

Do the same for shapes with one dot inside, and two dots, and so on.

35. FINDING TRIANGLES

Large equilateral triangles are made up from smaller ones. Investigate the number of different-sized triangles inside.

36. ADDING DIGITS

The digits of 122 are 1, 2, 2 and add up to 5.

Find all the numbers whose digits add up to 5.

37. ROUND AND ROUND

Choose any four numbers and place them at the corners of a square.

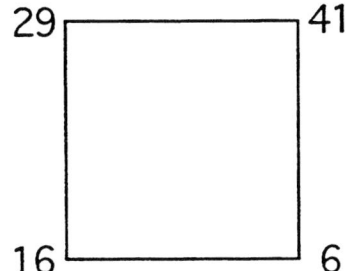

By the middle of each side of the square write the difference between the two numbers at the ends of that side.

Use these numbers for the corners of a new square and repeat the process.

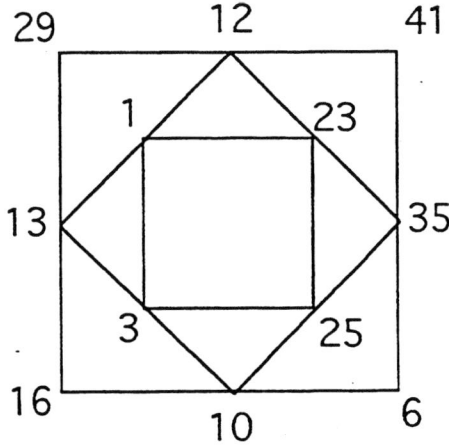

Investigate what happens.

38. ALL THE DIGITS

12 + 34 + 56 - 78 - 9 = 15

12 + 345 - 67 - 89 = 201

Keeping the digits 1 to 9 in order, what numbers can you make?

39. AFRICAN NETWORK PATTERNS

Use dotty paper.   These are the first 3 designs.

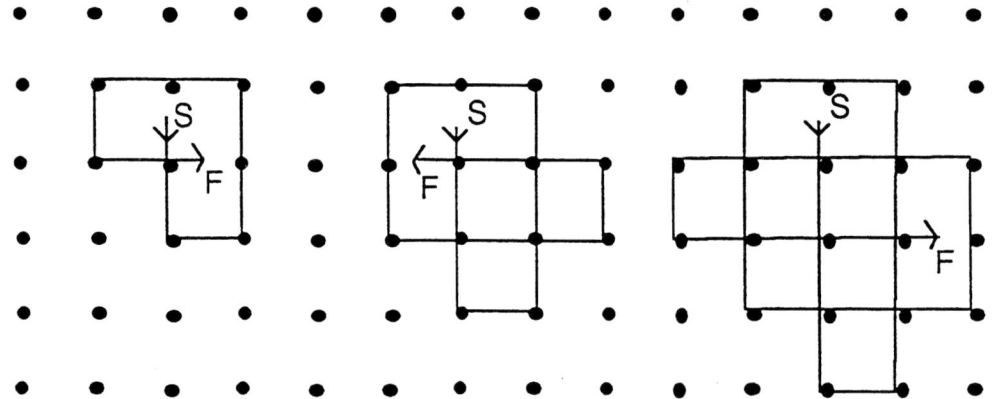

Each is drawn in a continuous movement.
Draw these to get the feel of how the patterns are created.

Draw some bigger ones.   What is the next size?

40. ELEPHANT WALK

An elephant, very fond of buns, walks through a set of cages each containing one bun.

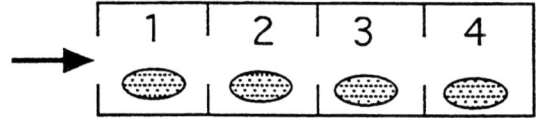

To get all of the buns the elephant must walk through a minimum of 7 cages. In this case it has to go back through 3 cages.

Try for these sets of cages:

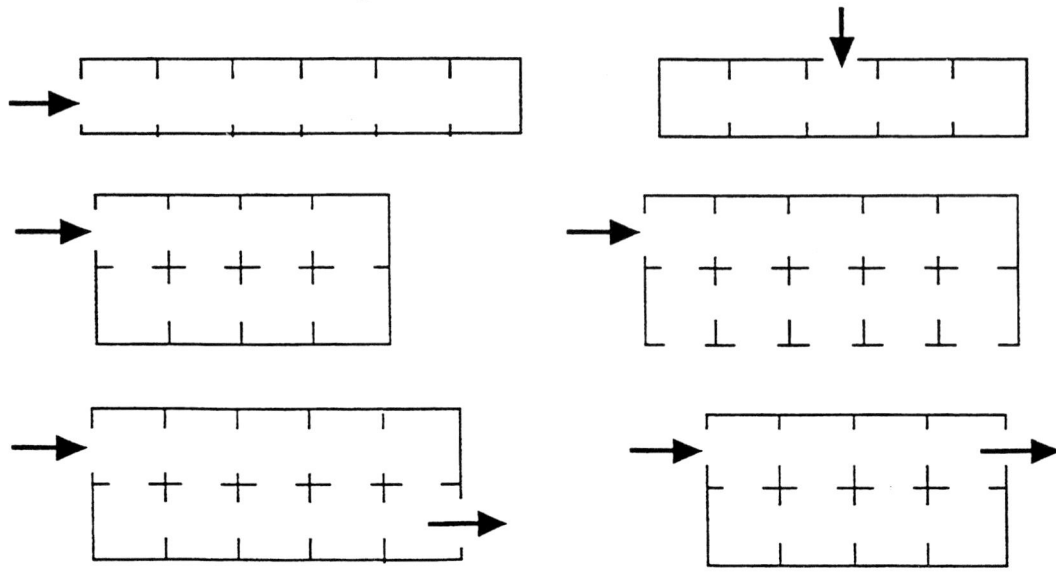

41. PALINDROMIC DATES

The 29th of November 1992 had a palindromic date:
29 - 11 - 92.
(It reads the same backwards as forwards).

When was the one before that?

When is the next palindromic date?

42. DOMINO ARITHMETIC

This is a domino multiplication:

```
  4 4 2
X     3
-------
1 3 2 6
```

Investigate other domino multiplications.

What other domino statements can you make?

43. CALENDARS

Start with a calendar and investigate row patterns and diagonal patterns.

How many months require 6 columns (or rows)?
How many in other years?

Look for Friday 13ths.
How many in different years?

44. TOTALS

Using the four numbers
$$1\quad 5\quad 8\quad 9$$
(each one once only)

and the symbols for the four basic operations

$$+\quad -\quad \times\quad \div$$

and brackets ( )

we can make various totals

for example $\quad 9 - 8 = 1$
$$(9 + 1) \div 5 = 2$$

What other numbers can you make?

45. PATIO PATHS

I have six metre-square paving slabs and I want to lay them to make a patio in my garden.

If I arrange them like this then I have 12 metres to trim round my patio.

If I arrange them like this then I have 14 metres to trim.

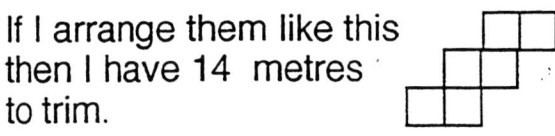

What is the shortest length of border I can have round my patio?

What is the longest?

46. THE TETHERED GOAT

A goat is tethered in a field.
The length of the tether is 9 metres.
What area of grass can the goat graze on?

Later the goat is tethered to the corner of a hut. If the hut is 4 metres by 6 metres, what area can the goat graze on now?

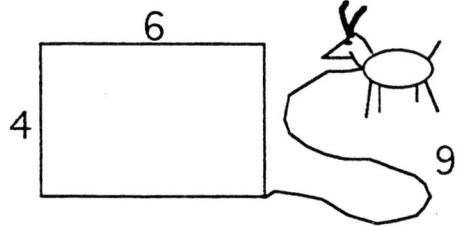

47. COUNTING TRIANGLES

The vertices of a square are joined in every possible way with straight lines.

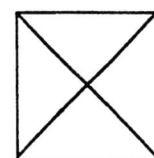

How many triangles are formed?
(large ones and small ones).

What about a regular pentagon?

48. TRAVELLING SALESPERSON

Choose four points on square dotty paper which can represent towns on a map.

Determine the shortest path that will join them all up.

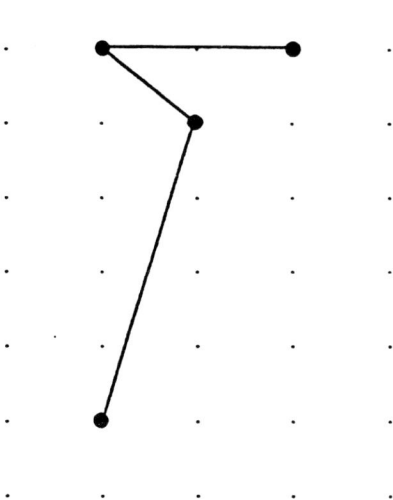

Try with different arrangements of dots.

49. NUMBER PATTERNS

Lay out 100 square tiling generators in a ten-by-ten square so that they all have the same orientation.

What happens if you give every third tile a quarter turn clockwise?

Every seventh tile?

What happens if you turn every third and then every fourth tile?

50. SUMS AND PRODUCTS

10 = 5 + 5   5 x 5 = 25

　 = 7 + 3   7 x 3 = 21

　 = 5 + 3 + 2   5 x 3 x 2 = 30

What is the greatest product that can be made from the numbers that add up to 10?

Try using a different starting number.

Is there a pattern?

51. TILINGS

Make a lot of square tiles with on half coloured and one half white (use one colour only).

What patterns can you make?

Try different ways of splitting the square.

52. PROJECTIONS

Cut a square from cardboard. View the square from different angles.

What shapes can you see?

53. STICKS

There are five different ways of connecting four sticks end to end:

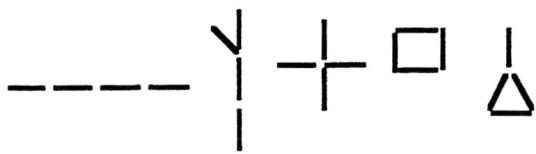

Investigate for different numbers of sticks.

54. RED AND YELLOW

How many different two-by-two-by-two cubes can you make using four red cubes and four yellow cubes?

55. ROUND THE BLOCK

Can you draw a line which starts at the black dot in the network, crosses each branch of the network once and only once, and returns to the black dot?

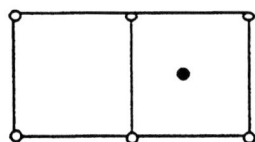

What about this one?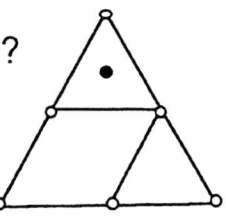

Experiment with other networks.

56. SUMS OF DIVISORS

Start with any number.
Map the number to the sum of its divisors, including 1, but excluding itself. Use this map repeatedly.

For example,
the divisors of 12 are 1, 2, 3, 4, and 6. 1 + 2 + 3 + 4 + 6 = 16

So, starting with 12 we get:

$$12 \longrightarrow 16 \longrightarrow 15 \longrightarrow 9 \longrightarrow 4 \longrightarrow 3 \longrightarrow 1 \circlearrowright$$

Investigate the properties of this mapping.

57. QUINCUNX

Quincunx is the name given to the traditional arrangement of five dots on dominoes.

The following growing pattern is based on the quincunx.

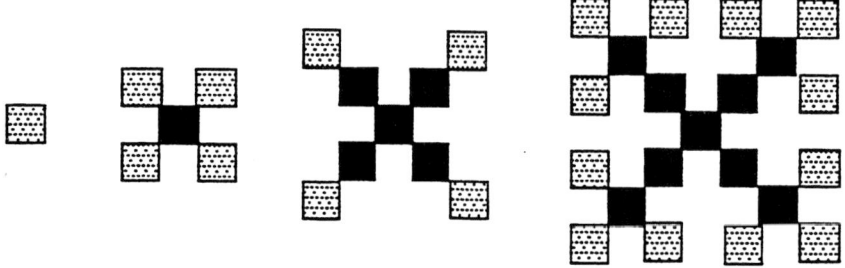

Copy the patterns on to squared paper and continue them.

How many squares are added to make the 5th pattern?
           the 6th pattern?
           the 7th pattern?
           the 8th pattern?

How many squares are there altogether in the 5th pattern?
          the 6th pattern?
          the 7th pattern?
          the 8th pattern?

58. STIFF LITTLE FINGERS

Each of the following diagrams is made up of four line segments on square grid. What others are there?

What about 5 line segments?

59. 10×12 OR 11×11?

Choose three consecutive numbers:

    10 11 12

Multiply the least number by the greatest:

    10 × 12

Multiply the other number by itself:

    11 × 11

Compare the results.

Try this with three more consecutive numbers.
Try several more times.
What happens?

60. FIRST WITH THE FACTORS

What is the first number with exactly three factors?

Exactly four factors?

Exactly five factors?

61. PATIO TILES

Cover this patio with tiles like this:

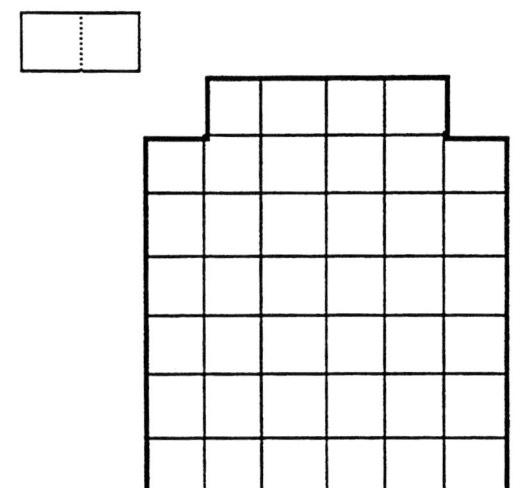

What about this one?

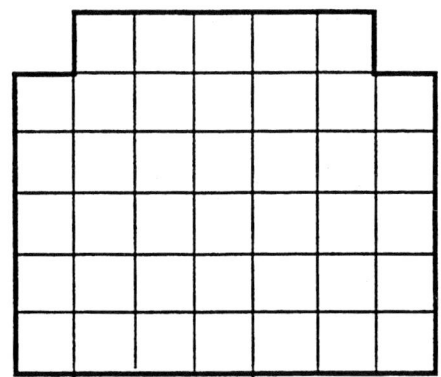

Try other shapes with an area of 40 squares.

62. LINES AND REGIONS

Draw 4 straight lines on a piece of plain paper so that you get the maximum number of crossing points. How many crossing points can you get?

How many inside regions are there?

63. EQUABLE RECTANGLES

Some rectangles have their perimeters numerically less than their area:

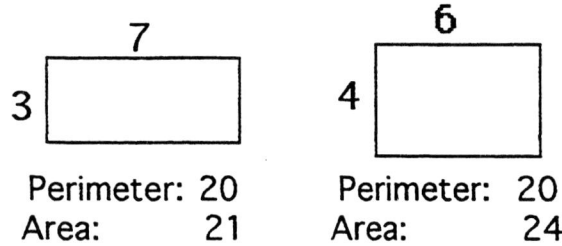

Perimeter: 20 Perimeter: 20
Area: 21 Area: 24

Can you find any rectangles whose perimeter is numerically greater than their area?

Can you find any squares or triangles whose perimeter is numerically equal to their area?

64. STICKY TRIANGLES

Using twelve sticks of equal length, what triangles can you make?

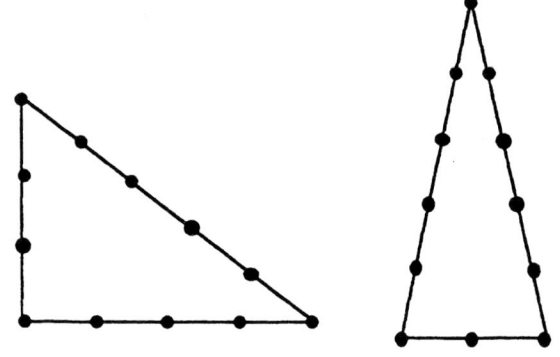

Try with other numbers of sticks.

Does the number of triangles depend on the number of sticks?

65. COLOURED CUBES

If the faces of a cube can only be coloured red, then you can make only one cube which will be red on each face.

How many different cubes can you make if each face can be either red or green?

66. THE GAME OF 25

The game is for two players and the rules are as follows:

The first player chooses one of the numbers 1, 2, 3, 4, 5, 6.

The second player chooses a number from the same set, and adds it to the first player's number.

The players continue to take it in turns to choose a number from the set and add it on to the previous total.

The player who makes the total up to 25 wins!

Can either player make sure of winning? If so, how?

67. A4

What is the largest container you can make from an ordinary A4 sheet of paper or card?

68. DIFFERENCING

Choose any two starting numbers. Make a chain using the rule:

Take the (positive) difference of the last two numbers to make the next term.
For example:

12, 5 —→ 7 —→ 2 —→ 5 —→ 3 ....

Investigate what happens.

69. NOUGHTS AND CROSSES

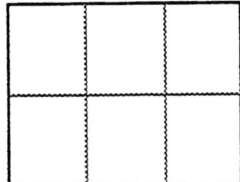

Can you devise a strategy for noughts and crosses that will ensure that you
    a) never lose?
    b) always win?

70. FOLDING STAMPS

Six postage stamps are in a block.

How many different ways can you find of folding them into one pile?

71. CUBOIDS

Find some cuboids that have the same volume but a different surface area.
For example:

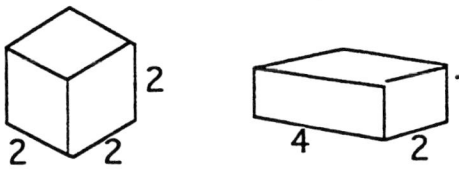

Volume: 8   Volume: 8
Surface area: 24   Surface area: 28

72. HOW GREAT?

Use the digits 1, 2, 3, 4, 5 exactly once each to make two or more numbers.

eg   4   21   53

Multiply these numbers together.

$$4 \times 21 \times 53 = 4452$$

Try other arrangements of the digits 1 to 5.

What is the greatest product that can be made?

73. BRICKS

Investigate different ways of tessellating two-by-one oblongs.

SOME NOTES AND EXTENSION IDEAS

1. ROUTES
Extension idea:
Invent your own grids and rules for moves.

3. DOTS AND LINES
Extension idea:
What happens if every dot must be joined to every other dot by a straight line?

4. MULTIPLICATION SQUARE
Extension ideas:
Make another and put in only the units.
Find the repeating patterns and invent a way of recording them.

5. FAULT LINES
Extension idea: Suppose you use 3 x 1 rectangles instead of dominoes?

6. DOMINOES
You will need to decide on the rules.
Do all the dominoes have to be used?

7. ALL DIFFERENT
Extension idea: What about seven dice? A good example of zero probability.

8. 1089
Extension idea: Are there similar results with 2 digits, 4 digits, 5 digits?

9. SIXES
Perhaps a discussion about the meaning of the word 'sum' is in order here.
To mathematicians it means 'add'. What do your pupils think?
Extension idea:
Try with other numbers such as ten.

10. SQUARES IN SQUARES
Extension idea:
How many squares on a chessboard?
Can you generalise the results for an n X n?
How many rectangles on any of the boards?

11. PARTITIONS
Interlocking cubes in two colours could also be used.

12. CUBE NETS
Extension idea:
What about nets for other solids?

13. PALINDROMES
Extension idea: The palindromic number 828 is the sum of 216 and its reverse 612.
Take some palindromic numbers.
Can you always make them by adding a number and its reverse?

15. CHAINS
Try to put all your results together on one diagram.
Extension idea: Try changing the rules e.g. alter Rule 2:   If the number is odd, multiply it by 3 and subtract 1.

17. MAX BOX
Start with plenty of scrap paper, scissors and, for younger pupils, some cubes for measuring the volume.
Extension idea: Try with a rectangular piece of card, say, 10cm X 20 cm.

18. FOUR FOURS
Square roots could perhaps be used by those who know about them.

$$4 + 4 + \frac{4}{\sqrt{4}} = 10$$

19. PAINTED CUBES
Extension idea:
Investigate this for different sizes of cube.

20. CONSECUTIVE SUMS
Powers of two cannot be split up into consecutive sums.

21. SETS OF FIVE
Extension ideas:
Try to make a set of numbers which produces every number up to the highest.
Can you make a set which will give every number from 1 to 20 but none above 20?

22. ARITHMOGONS
Arithmogons can be in the form of squares or other polygons. If only the numbers in the middle squares are given then at this stage they can only be solved by trial and improve so the numbers should be small.

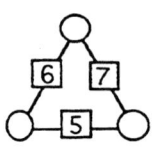

23. MAGIC SHAPES
The numbers from 1 - 10 and 1 - 12 are needed for these shapes:

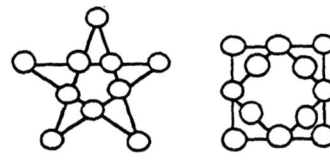

4. HALVING THE BOARD
Extension idea:
Try other boards such as a 10 x 20 rectangle.

25. HAPPY NUMBERS
Extension ideas: Is there a quick way of telling if a number is happy?
Find a way of recording all the happy and 'sad' numbers under 100.

26. TABLES
The third example requires the positive *difference*. This might lead to a useful discussion about this difficult word.
The third incomplete table can be completed in many ways.
Extension idea:
Make up your own puzzles of this kind.

27. MOVING TESSELLATIONS
Plenty of space and a box of ATM MATs are really needed for this starting point.

29. RECTANGLE AREAS
Extension idea:
Try with triangles on isometric dotty paper.

30. STRIPS OF SQUARES
Use linking cubes to start with. Use squared paper for recording.
Extension idea: Investigate for different strip lengths and different numbers of colours.

32. POLYGON SYMMETRIES
Extension idea: What about hexagons?

33. HEXIAMONDS
The name is made up from the word 'diamond.' A diamond is, of course, made up from two triangles. There are also 'triamonds' etc.
Extension idea: Investigate the way that hexiamonds can tessellate.

35. FINDING TRIANGLES
Isometric dotty or lined paper would be helpful, or even a set of plastic equilateral triangles.

36. ADDING DIGITS
Extension idea: Investigate numbers whose digits add to other totals.

37. ROUND AND ROUND
Extension idea:
Use a triangle, or a pentagon or ....

40. ELEPHANT WALK
Extension idea:
Investigate for different sets of cages with entrances and exits in various places.

42. DOMINO ARITHMETIC
Extension idea: Try with a set 0 - 9 dominoes.

43. CALENDARS
Extension ideas: What is the probability:
that the 13th will be a Friday?
that the first day of the month, year, century will be a Friday?
of 5 Fridays in a month?
of 5 Fridays in February?

44. TOTALS
Extension ideas:
Find the longest run of totals 1, 2, 3, ... that can be made without any omissions.
Investigate for different sets of four numbers.

45. PATIO PATHS
Extension idea: What if you use different numbers of paving slabs?

46. THE TETHERED GOAT
Extension idea: Investigate for different lengths of tether, and sizes and shapes of hut.

47. COUNTING TRIANGLES
The square has 4 large and 4 small triangles.
The pentagon 5 each of 5 different triangles.

48. TRAVELLING SALESPERSON
Extension idea:
What about joining more than four points?

49. NUMBER PATTERNS
Extension idea: What happens if you use a different size of square layout, or a rectangle?

50. SUMS AND PRODUCTS
One approach might be to first break up 10 in as many ways as can be found, then replace the plus signs with multiplication signs and work out the answers.

51. TILINGS
Extension idea:
Try two-by-one rectangles instead of squares.

52. PROJECTIONS
Extension idea: Investigate views of other shapes such as a triangle, a circle, a cube.

53. STICKS
It will be necessary to discuss why there are no more ways. The first example below can be rearranged into the second without changing where the sticks are connected.

Extension idea:
Explore other situations. For example, here are the 5 possible '3-plants' (each has one point anchored to the 'ground').

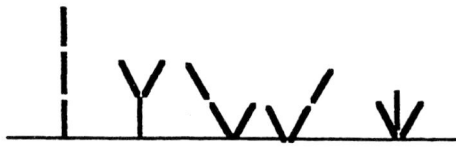

54. RED AND YELLOW
Extension ideas:
Or with three red cubes and five yellow?
Or with three colours?

55. ROUND THE BLOCK

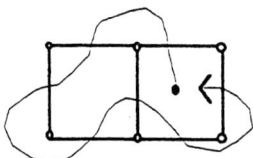

Extension idea: Does changing the position of the black dot effect the result?

57. QUINCUNX
To follow the rule the squares added at the corners must not touch.

	Added	Total
4th pattern:	12	21
5th pattern:	4	25
6th pattern:	12	37
7th pattern:	12	49
8th pattern:	48	97

59. 10 X 12 OR 11 X 11?
Extension ideas: Why does this happen?
Try again, but with numbers that are increasing by 2 instead of 1.
e.g. compare 10 X 14 with 12 X 12.

62. LINES AND REGIONS
Extension idea:
Investigate for other numbers of lines.

63. EQUABLE RECTANGLES
If the sides are less than 4 units, then the perimeter will be numerically less than the area. A 4 x 4 square has equal perimeter and area.
Extension idea: Try comparing surface area of solids with the volume.

64. STICKY TRIANGLES
Younger pupils can try with a constructional kit such a 'Geostrips'.
Extension idea: What triangles can you make with a length of string?

65. COLOURED CUBES
Try making the whole set of different cubes. You will need to discuss rotation and reflection.
Extension ideas: What happens if you can use three colours? What about tetrahedrons?

67. A4
Some sticky tape will be needed! Some dry sand might help younger pupils measure the volume, but card, rather than paper should be used in this case.

69. NOUGHTS AND CROSSES
Extension idea: The rules of the game could be changed. For example, the players could play a nought *or* a cross on their move and the first player to complete a line of three noughts *or* three crosses would be the winner.

70. FOLDING STAMPS
A method of recording will be needed. one way would be to number each stamp in the folded pile. For example:

5	4	1
6	3	2

71. CUBOIDS
With younger pupils try using interlocking cubes.
Extension idea: Can you now find some cuboids which have the same surface area but different volumes? (This is hard!)